Biodiversity Series of Chongming Dongtan
崇明东滩生物多样性科普丛书

无形的联系

大滨鹬的故事

Invisible connection
a story of Great Knot

陈婷媛 文
吴巍 译
王紫 绘

上海科学技术出版社

序

在全球的一万多种鸟类之中，接近四分之一的种类是具有迁徙习性的候鸟。它们每年在相隔数千千米乃至上万千米的繁殖地和越冬地之间往返飞行。鸟类迁徙是地球上最具吸引力的自然现象之一。迁徙活动促进了全球不同区域之间的物质流、能量流和信息流，对于维持地球系统的生态过程和生态平衡起着重要作用。

许多候鸟的迁徙路程遥远，它们在迁徙途中需要有一个或多个迁徙驿站，以便进行休息和能量补给，为下一阶段的迁徙飞行做准备。位于我国长江口的崇明东滩就是东亚－澳大利西亚候鸟迁飞区的一处重要迁徙驿站。崇明东滩的迁徙候鸟种类多、数量大，其中鸻鹬类是崇明东滩最具代表性的候鸟。它们的繁殖地主要位于俄罗斯西伯利亚和美国阿拉斯加的高纬度地区，越冬地主要位于澳大利亚、新西兰及东南亚地区。崇明东滩每年迁徙过境的鸻鹬类数量达数十万只，在迁徙鸻鹬类的保护方面起着关键作用。

本书采用精美的手绘图和简洁的文字，向读者描绘了崇明东滩的一种常见鸻鹬类——大滨鹬的迁徙活动。大滨鹬体重只有100多克，但它是全球20多种滨鹬类中体型最大的种类，因而得名"大"滨鹬。大滨鹬是东亚－澳大利西亚迁飞区的特有鸟类，它们繁殖于俄罗斯远东地区，主要在澳大利亚北部越冬，在春季和秋季迁徙时于崇明东滩停歇。除了夏季繁殖期之外，大滨鹬在其他季节只在沿海地区活动。它们主要在滩涂湿地觅食，因此对滩涂湿地的环境变化非常敏感。近几十年来，由于东亚沿海地区的滩涂湿地面积减少、质量下降，大滨鹬的种群数量快速下降。目前，大滨鹬被列为我国二级重点保护鸟类，被世界自然保护联盟列为"濒危"物种。

为了了解大滨鹬等鸻鹬类的迁徙活动，崇明东滩自然保护区的工作人员从20世纪80年代开始采用环志等方法研究鸟类的迁徙活动。在东亚－澳大利西亚候鸟迁飞区还有许多国家和地区的研究人员也在开展鸻鹬类环志。研究人员追随着鸟类的迁徙路线，在候鸟的繁殖地、迁徙停歇地和越冬地开展调查研究。通过不同地区研究人员的合作与交流，掌握了鸻鹬类的迁徙活动规律，为开展鸟类保护工作提供了重要资料。书中的"白胡子爷爷"便是这些研究人员的代表。

读者可以从这本书中初步了解鸟类研究人员和候鸟的故事。希望本书读者在未来能够成为鸟类保护的参与者。

2023年3月1日于复旦大学江湾校区

Foreword

Among more than 10,000 species of birds in the world, nearly a quarter are migratory birds. They fly back and forth between breeding grounds and wintering grounds that are thousands or even tens of thousands of kilometers apart every year. Bird migration is one of the most attractive natural phenomena on Earth. Migration activities promote the flow of matter, energy and information between different regions of the world, and play an important role in maintaining the ecological processes and balance of the Earth system.

Many migratory birds have long migration routes and need one or more migration stopovers along the way to rest and replenish energy for the next stage of migration. Chongming Dongtan, located at the estuary of China's Yangtze River, is an important migration stopover site in the East Asian-Australasian Flyway. Chongming Dongtan has a large number and variety of migratory birds, among which sandpipers are the most representative. Their breeding grounds are mainly located in high-latitude areas such as Siberia in Russia and Alaska in the United States, and their wintering grounds are mainly in Australia, New Zealand and Southeast Asia. Hundreds of thousands of sandpipers pass through Chongming Dongtan every year during migration. Chongming Dongtan is playing a key role in the conservation of migratory sandpipers.

This book uses beautiful hand-drawn illustrations and concise texts to depict the migration activities of a common sandpiper species on Chongming Dongtan - Great Knots. A Great knot weighs only about 100 grams, but it is the largest species among more than 20 species of knots worldwide, hence their name "Great" Knots. Great Knots are endemic birds to the East Asian-Australasian Flyway. They breed in Far Eastern Russia and mainly winter in northern Australia. They stop over at Chongming Dongtan during spring and autumn migrations. Except for summer breeding season, Great Knots only live in coastal areas. They mainly feed on tidal flats, so they are very sensitive to environmental changes on tidal flats. In recent decades, due to the reduction and degradation of tidal flats along East Asia's coastlines, Great Knots' population has declined rapidly. Currently Great Knots are listed as second-class state protected wildlife in China and as an "endangered" species by IUCN.

In order to understand the migration activities of sandpipers such as Great Knots, researchers of Chongmirg Dongtan National Nature Reserve began to study bird migration by methods such as banding since 1980s. There are also many researchers from other countries and regions who are conducting banding studies on sandpipers in East Asian-Australasian Flyway. Researchers follow bird migrations and conduct surveys at breeding grounds, stopover sites and wintering grounds for migratory birds. Through cooperation and communication, researchers from different regions have mastered some rules about sandpiper migrations which provide important data for bird conservation work. The "Grandpa Melville" in this book is a representative of these researchers. Readers can learn some stories about bird researchers and migratory birds from this book. I hope readers would also become participants in bird conservation work someday in the future.

Zhijun Ma
March 1st, 2023 at Fudan University, Jiangwan Campus

前言

"小燕子，穿花衣，年年春天来这里"，这首我们耳熟能详的儿歌《小燕子》展现了燕子迁徙的景象。鸟类迁徙是大自然中非常普遍的自然现象，每年都有数十亿的候鸟往返于繁殖地和越冬地。

经过研究，科学家将全球鸟类迁飞区划分成九个，上海崇明东滩就处在东亚－澳大利西亚这个迁飞区。东亚－澳大利西亚迁飞区覆盖了22个国家和地区，是全球最大的迁飞区。每年春天有超过几千万只水鸟从澳大利亚和新西兰等地起飞，跨越大洋来到中国，其中大部分鸻鹬类在沿海的泥滩上觅食补充能量，比如长江口、黄河口、鸭绿江口。短暂停歇后它们再次起飞，飞往俄罗斯西伯利亚和美国阿拉斯加进行繁殖，繁殖结束后再返回澳大利亚和新西兰等地越冬。所以对于大多数鸻鹬类而言，西伯利亚和阿拉斯加是繁殖地，澳大利亚和新西兰等地是越冬地，而在中间做短暂停歇的地方就是中途停歇地。迁徙期间，这些水鸟要依赖一系列优质湿地进行休息和觅食，积聚充足的能量，完成下一阶段旅程。如果把鸟类迁徙路线比喻成一条高速公路，那中途停歇地就是重要的服务区。崇明东滩的地理位置与自然条件得天独厚，是东亚－澳大利西亚迁飞路线的重要组成部分，也是鸟儿迁徙这条高速公路上非常重要的服务区。

从人类的视角来看，迁飞区内的各个国家和地区都有着基于历史、商业、条约和协议方面的重要联系，而通常没有考虑国家和地区间的生态联系。正是因为迁徙鸟类全球性的奇妙飞行，把不同思想、不同语言和不同文化，甚至相隔数千千米的国家联系在一起，共同开展科学研究，促进全球生态的保护与合作。正是迁徙路线上许许多多关心鸻鹬类保护的人们一起携手共进，推动了鸻鹬类的保护事业。在此，向所有关心和爱护鸟类的朋友们表达深深的敬意。

Preface

"Little swallow, wearing a beautiful dress, comes here every spring", this familiar Chinese children's song *Little Swallow* shows the scene of swallows' migration. Bird migration is a very common natural phenomenon. Every year there are billions of birds traveling between breeding grounds and wintering grounds.

Through research, scientists have divided the global bird migration routes into nine flyways. Chongming Dongtan in Shanghai, China is located in the East Asia-Australia flyway. The East Asia-Australia flyway covers 22 countries and regions, It is the largest flyway in the world. Every spring, tens of millions of water birds take off from Australia and New Zealand and fly across the ocean to China. Among them, the waders are a very important group. Most of the waders feed on mudflats along the coast, such as the Yangtze River estuary, Yellow River estuary, Yalu River estuary etc. After a short stopover, they take off again and fly to Siberia in Russia and Alaska in the United States for breeding. After breeding, they return to Australia and New Zealand for wintering. Therefore, for most waders, Siberia and Alaska are breeding grounds, Australia and New Zealand are wintering grounds, and places where they make short stopovers are stopover sites. During migration, these water birds rely on a series of high-quality wetlands for rest and foraging, accumulating enough energy to reach the next stage of their journey. If we compare the bird migration routes to highways, then the stopover sites are just like service areas. Chongming Dongtan has a unique geographical location and natural conditions. It is a very important service area on this bird migration highway.

From human's perspective, each country and region within the flyway has important connections based on history, commerce, treaties and agreements, but usually the ecological connections are ignored. It is precisely because of these amazing flights of migratory birds that these countries

《无形的联系——大滨鹬的故事》一书就是以鸟类迁徙作为故事的背景，并以"大斌"和"小玉"两位主人公的奇妙迁徙之旅，串联起了鸟类越冬地、中途停歇地和繁殖地之间的故事。遍布迁徙路线上的各种类型的栖息地是鸟类赖以生存的重要生态空间，在全球生态环境急剧变化的过程中，对栖息地的保护显得尤为重要，需要我们共同努力。作为一名鸟类保护工作者，普及鸟类科学知识是我的职责所在，通过绘本这种形式讲述鸟类迁徙、鸟类栖息地保护是我继《盼归栖——小天鹅的故事》一书以来的第二次尝试，希望有助于读者进一步了解鸟类保护生物学的知识和我们保护工作者所做的工作以及所付出的努力，也希望有更多的读者积极参与到鸟类保护工作中来。

　　常言道"燕雀安知鸿鹄之志"，其实无论是燕雀还是大雁、天鹅，都是天空中南来北往、纵横万里的生灵，是全球生物网络的重要部分，它们的迁徙飞行，是自然界最美妙的画卷，也是对生命的最高礼赞！

陈婷媛

2022 年 12 月 16 日于上海崇明东滩

and regions with different ideas, languages and cultures are connected together, although they might be thousands of kilometers apart. They carry out scientific research and conservation cooperation together. Because of the joint effort of many people who care about migratory waders' conservation along this flyway, the migratory waders' conservation is pushed forward. Here, I would like to express my deep respect to all friends who care and love these birds.

The book *Invisible connection-a story of Great Knot* uses bird migration as the background. Through the wonderful journey of two protagonists "Dabin" and "Xiaoyu", the stories happened in wintering grounds, stopover sites and breeding grounds are connected. Various habitats along this migration route are important ecological spaces that birds rely on. With the rapid change of global ecological environment, the protection of habitats becomes particularly important. It requires our joint efforts. As a bird conservation worker, to popularize scientific knowledge about bird conservation is my duty. To tell stories about bird migration, bird habitat and conservation through picture books is my second attempt since my last book *Longing for Your Return-A Story of Tundra Swan*. I hope it could help readers further understand knowledge about bird conservation and our efforts. I also hope more readers could actively participate in bird and wetland conservation.

Whether sparrows, geese or swans, they are all creatures flying over thousands of miles between south and north. They are all important parts of the global ecosystem. Their migratory flight is the most wonderful painting of this nature and the highest praise of life!

Tingyuan Chen
December 16th, 2022 at Shanghai Chongming Dongtan

It is February on the beach of Broome, Australia. A large flock of Great Knots are taking their final energy supply. Soon, they will set off for the far northern hemisphere and start another year's journey for breeding. This journey is dangerous but romantic. For these birds, this long-distance migration is not a burden, but a wonderful journey for food and love.

2月，澳大利亚布鲁姆的沙滩上，一群越冬的大滨鹬正在进行能量补给，为去往繁殖地积攒脂肪。再过不久，他们就要启程奔赴遥远的北半球，开始又一年的繁衍旅程。这段旅程既艰险又浪漫，对他们来说，长距离的迁徙不是负担，而是奔赴美食和爱情的奇妙之旅。

滩涂上丰富的底栖动物，使这里成为鸻鹬类休憩觅食的天堂。每当潮水退去，饥饿的鸻鹬们就立马来到滩涂上享用大餐。

在这片清澈细腻的沙滩上，经常能看到一位长者的身影，大家都亲切地称他为"白胡子爷爷"。和其他从世界各地赶来的国际志愿者一样，白胡子爷爷是一位热衷于鸟类保护的专家。

Abundant benthic animals on the tidal flat make it a paradise for shorebirds to rest and forage. Whenever the tide goes down, hungry shorebirds come to the tidal flat to enjoy a big meal.

In here you can often see an old man, whom everyone affectionately calls "Grandpa Melville". Like other international volunteers from all over the world, "Grandpa Melville" is also passionate about bird conservation.

为了研究鸟类迁徙的奥秘,每年2月,研究者会用炮网捕捉那些正在休憩觅食的鸻鹬类。用火药射出的巨大捕鸟网最多一次能捕捉到300~400只鸻鹬类,炮网发射后,志愿者们一拥而上,快速地从网里取出鸟儿,为他们做测量、体检,并佩戴金属环和旗标。

做完环志的鸟儿,就有了鲜明的彩色标记。当鸟儿在迁徙时,各国的志愿者们就可以在停歇地观察到这些带着标记的鸟儿,从而了解鸟儿们都在哪里被环志过。

To study the mystery of bird migration, researchers use cannon nets every February to catch shorebirds that are wintering in Broome. The huge bird-catching net fired with gunpowder can catch 300-400 shorebirds each time at the most. After the net is launched, volunteers swarm up to quickly take the birds out of the net, measure them, perform physical examinations, and put metal rings and leg flags on their legs.

When the banding is finished, every bird will carry a distinctive colour leg flag. Volunteers from various countries could easily recognize the marks when these birds are migrating in the flyway and know where they have been banded.

This female Great Knot with a yellow flag on the leg is "Xiao Yu". She and her companions have had enough food, and they are about to start their journey to the north. Their destination is the distant Siberia. During the journey, they will fly over tens of thousands of kilometers, overcome many difficulties, and finally reach the mysterious land of love to breed.

这只脚上戴着黄色旗标的大滨鹬名叫"小玉"。此时,她和同伴们已经吃饱喝足,就要开启北迁之旅了,目的地是遥远的西伯利亚。他们将飞越上万千米,克服重重困难,到达神秘的爱情圣地繁衍后代。

飞过无边的海洋，越过美丽的海岛，经过不眠不休的长距离飞行，体力即将耗尽的他们，急需找一个地方休息并补充食物。

看，远处那一片绿洲是哪儿？原来，这就是上海的崇明岛！是父辈们口口相传的"迁飞高速公路"上非常重要的服务区和补给点。

Flying over the boundless sea and beautiful islands, after this long and sleepless flight, they are in urgent need of finding a place to rest and replenish their energy.

Look, where is the oasis in the distance? It is Chongming Island in Shanghai, China! It is a very important service area and supply station on the "Migration Expressway". This information has been passed on from generation to generation.

对大滨鹬而言，上海崇明岛东部的滩涂是一片神奇而美丽的湿地。往年迁徙时，不少大滨鹬都来过这里，优美的环境和丰盛的食物，给他们留下了美好的回忆。如今故地重游，看到熟悉的滩涂、肥美的吃食和善良的人们，大滨鹬们感到分外亲切。

For Great Knots, the tidal flat of Chongming Dongtan is a magical and beautiful wetland. Many Great Knots have been here during their migration in the past. The beautiful environment and rich food left them with very good memories. Now seeing the familiar tidal flat, plentiful food and kind people again, the Great Knots feel very comfortable.

广袤的滩涂上，还有许多其他迁徙到此休憩的伙伴，比如黑脸琵鹭、勺嘴鹬、大杓鹬和灰斑鸻等。琵鹭用他们扁平如汤匙的长嘴捕捉鱼类。鸻鹬类则用尖尖的嘴巴捕食底栖动物，他们将贝壳等底栖动物整个吞下，并在砂囊中碾碎。底栖动物通常会待在一个地方，当鸻鹬类发现了一片食物丰富的区域时，众多同类都会聚集过来一起觅食。

瞧！一只大杓鹬正用长长的喙，牢牢钳住一只螃蟹猛烈抖动。他打算甩断螃蟹危险的蟹爪，然后再将它整个吞下。这是他们典型的捕食方式。

On the vast tidal flat, there are many other birds, such as Black-Faced Spoonbills, Spoon-Billed Sandpipers, Far Eastern Curlews and Grey Plovers. Spoonbills use their beaks, which are flat like spoons, to catch fish; shorebirds use their sharp-pointed bills to prey on benthic animals. They swallow shellfishes whole and crush them in gizzards. Benthic animals usually stay in one place. When shorebirds find a place rich in food, many of them will gather to forage.

A Far Eastern Curlew is holding a crab firmly with its long beak and shaking it violently. He is trying to snap off the crab's dangerous claws before swallowing it.

某天清晨，崇明东滩保护区的空气特别清新。一只名叫"大斌"的大滨鹬，正在自由翱翔。这时，大斌听见了同伴的叫声，便缓缓扇动翅膀飞了过来，刚准备降落，他突然一头撞进了一张大网。

大斌瞬间被恐惧笼罩。

One fresh morning, a Great Knot named "Da Bin" is flying freely. Da Bin hears the song of another Great Knot, so he slowly approaches. Just as he is about to land, he suddenly slams into a large net.
Da Bin is instantly covered in fear.

他被装进一只大箩筐中，发现里面还有一位美丽的姑娘。"嗨，你好，我叫大斌。你是怎么被关进这里的？""你好，我叫小玉。我是不小心撞进网里的。人类会不会吃了我们呀？"两个瑟瑟发抖的小家伙挤作了一团。

He is put into a big basket and finds a beautiful girl next to him. "Hello, my name is Da Bin. How did you get locked up here?" "Hi, my name is Xiao Yu. I accidentally fell into the net. Will human eat us?" Two little shivering birds huddle together.

随后，大斌和小玉被带进了一个房间。在那里，他们竟然看到了布鲁姆的白胡子爷爷！小玉终于放心了，安慰大斌说这些人并不是来吃他们的。白胡子爷爷为他俩做了全身体检，测量翅膀、称体重，还给大斌戴上金属环和上黑下白的旗标。

Later on, Da Bin and Xiao Yu are brought into a room. There, they see Broome's Grandpa Melville! Xiao Yu finally feels relieved and comforts Da Bin that these people will not eat them. Grandpa Melville does a physical examination for them, measures their wings, weighs them, and puts a shiny metal ring and black and white leg flags on Da Bin.

做完体检,大斌和小玉被爷爷放飞,回到了崇明东滩的茫茫滩涂。他们毫发无伤,只不过大斌腿上多了金属环和上黑下白的旗标。

After the examination, Da Bin and Xiao Yu are released by Grandpa Melville and return to the tidal flat. They are unscathed, except that there are a metal ring and black and white flags on Da Bin's legs now.

原来，刚才是保护区的工作人员和国际志愿者们在开展环志工作，真是虚惊一场！大斌和小玉一边听同伴们讲述保护区开展的鸟类研究监测工作，一边抓紧机会吃东西。忽然，大斌发现了一只个头超大、肉质鲜美的贝壳，他担心瘦弱的小玉不能和他一起并肩飞到繁殖地，便贴心地把贝壳衔给了她。

It turns out to be a banding research project carried by the staff of the Reserve and international volunteers. Da Bin and Xiao Yu seize the opportunity to eat while listening to their companions talking about the bird research and monitoring work carried out in the Reserve. Suddenly, Da Bin finds a huge shell with delicious meat. He gives the shell to Xiao Yu so that she can have enough energy to fly to the breeding ground with him.

　　在东滩的逗留，短暂而开心。大斌和小玉大部分时间都在生态修复区里的漫滩岛屿上休憩，或者到堤外滩涂上觅食。每天，他们都会看到工作人员在附近巡逻，为他们打造安全的避风港湾。保护区为鸟儿们提供了安全的环境，让他们住得好、吃得好，为即将到来的长距离飞行储蓄能量。

The stay in Dongtan is short but pleasant. Da Bin and Xiao Yu spend most of their time resting on the islands in the restoration area or looking for food on the tidal flat outside the seawall. Every day, they see staff patrolling nearby to protect them. The Reserve provides a safe environment for birds, allowing them to live well, eat well, and store energy for the upcoming long-distance flight.

远行的日子，终于到了。大滨鹬们将再度飞行6000多千米，途经黄海北部区域并在那里进一步积累能量，最终到达遥远的西伯利亚，在那儿繁衍后代。不料，刚起飞不久，他们就遭遇了暴风骤雨的恶劣天气。一个闪电劈过，大斌和小玉被迫分开了。

The day of departure finally comes. Great Knots are about to fly 6,000 kilometers to the remote Chukchi Peninsula in Siberia, where they will breed. Unexpectedly, shortly after takeoff, they encounter a violent storm. The lightning strikes. Da Bin and Xiao Yu are forced to separate.

大斌降落的地方，是一片沿海滩涂。这里满目疮痍，遍地垃圾，赖以生存的栖息地被渐渐蚕食。远处，工业废气和尘土弥漫整个天空，将天空染成一片灰黄。大斌不由揪起了心："不知道小玉现在怎样了？"

The place where Da Bin lands is another tidal flat. It is devastated and littered. In the distance, industrial emissions and dust filled the sky, turning the sky into dark grey. Da Bin can't help worrying, "How is Xiao Yu doing now?"

大斌尝试着换了一个地方,但情况也不容乐观。渔民在滩涂上作业,捕捞鸻鹬类喜欢吃的螃蟹和贝壳。

Da Bin tries to find another place, but the situation is still not good. In the new place, fishermen work on the tidal flat catching crabs and shells that shorebirds like eating.

不过，在鸟类保护者们的努力下，整个黄渤海地区的滨海岸线上，建立起了许多自然保护区。科研人员开展了鸟类迁徙的研究和保护，让更多人了解鸟类迁徙的奥秘。

However, thanks to the efforts of bird conservationists, many nature reserves have been established along the coastline of the Yellow and Bohai Sea. Researchers have carried out many research works about bird migration and conservation, so that more people can understand the mystery of birds.

　　一次次观鸟科普活动，让人们了解了鸟类的故事。保护区一系列形式多样的活动，让越来越多的人爱上了这些在天空中自由翱翔的小精灵，都自觉加入保护鸟类的行动中。

Bird watching activities let people know the story of birds. A series of various activities in the Reserve have made more and more people fall in love with these little spirits that fly freely in the sky, and consciously join in the action of birds' conservation.

43

经过长途跋涉，飞越千山万水，鸟儿们终于抵达了北迁的最后一站——西伯利亚的楚科奇。当第一批鹬鹬类从黄海抵达时，这里的冻原还没有完全融化，一片千里冰封的自然风光。

After a long journey flying over thousands of mountains and coasts, the Great Knots finally arrive at the destination of their northward migration-Chukchi Peninsula in Siberia. When the first flock of Great Knots from the Yellow Sea arrive, the snow has not completely melted yet. Part of the tundra and the mountains are still covered by snow.

不久，寒冷的北极迎来了冰雪消融。来自不同地方的鸟儿们，汇聚在这片湿地上，宛如一场热闹的"国际聚会"。红腹滨鹬和斑尾塍鹬从澳大利亚和新西兰涌向西伯利亚东北部，在中国华南部分地区和东南亚越冬的勺嘴鹬也抵达了。最令人欣喜的是，大斌和小玉也在这里浪漫相逢了。

Soon, the cold Arctic ushers in melting ice and snow. Birds from different places gather in this wetland, like a lively "international party". Red Knots and Bar-Tailed Godwits flock from Australia and New Zealand to northeastern Siberia, and Spoon-Billed Sandpipers that winter in parts of southern China and southeast Asia also arrive. The most gratifying thing is that Da Bin and Xiao Yu also have a romantic reunion here.

The Great Knots begin to establish their own territory, build nests, and breed. Their nests look simple and shallow. They are made of lichen, moss and leaves, and covered by sedges. It has been almost a month since they arrived here. On this day, a "twittering" sound suddenly appears on the tundra. A little life is born! Subsequently, many new lives come one after another, and various sounds of baby knots bring vitality to the tundra in early summer.

他们开始建立自己的领地，筑巢、繁殖。巢的构造很简单，浅浅的，由地衣、苔藓、树叶筑成，被遮蔽在薄薄的、相互交织的莎草下。迁徙的鹬鹬类到达北极，已经快一个月了。这天，冻原的上空突然出现了一阵"叽叽喳喳"的声音，一个小生命诞生了！随后，一大批新生命接踵而至，各种"咕咕嘎嘎"的声音，为初夏的北极带来了勃勃生机。

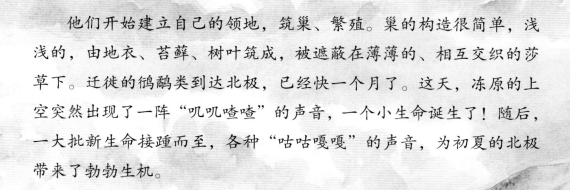

大斌和小玉也孕育了三个孩子。雏鸟们长着奇特的大脚，出生刚几个小时，就出巢开始探索世界了。冰雪融化后，无数的昆虫和浆果成了他们的美餐。不过他们也有需要警惕的敌人，那就是北极狐、贼鸥和棕熊。

Da Bin and Xiao Yu also have three children. The chicks have big feet as their parents. In just a few hours after they were born, they leave the nest and start exploring the world. As the ice and snow melted, countless insects and berries become their delicious meals. But they also have enemies to be wary of, Arctic Foxes, Skuas and Brown Bears.

又过了3周左右，雏鸟们开始了第一次飞行。不久后，他们的父母便要离开他们，飞往最近的海湾河口，为迁徙做准备。到那时，孩子们将开始独立生活，他们仍然会逗留在冻原和附近的内陆湿地，为自己的第一次长距离迁徙做准备。

After another three weeks or so, chicks start their first flight. Soon, their parents will leave them and fly to the nearest estuary to prepare for the southward migration. By then, these little birds will have to live independently, and they will still linger on the tundra and nearby inland wetlands, preparing for their first long-distance migration.

孩子们每天都在刻苦地练习飞行，胆子越来越大，肌肉也越来越结实。当夏末来临时，来自基因深处的本能向他们发出了迁飞的召唤，这一声声召唤是如此急切、如此神秘、如此不可抗拒。

万事俱备，终于迎来了开启伟大迁徙之旅的时刻。他们用力挥动着翅膀，奋力飞向前方，直到把地平线远远地甩到了身后。而深爱着他们的白胡子爷爷，则一直站在那儿，目送着他们离开。

Baby knots are practicing flying hard every day. Their feathers are now fully grown. Their hearts are getting stronger as well as their muscles. When the end of the summer comes, their instinct sends out a call to them to start migration. This call is so urgent, so mysterious, and so irresistible.

Everything is ready. Finally the moment to start the great migration arrives. They flap their wings vigorously and fly forward until they leave the horizon far behind. Grandpa Melville, who loves them deeply, stands there all the time, watching them leave.

　　在这次旅程中，大滨鹬们要飞越大半个地球，抵达西北澳这片辽阔的滩涂湿地。中途，他们还会造访父母曾去过的中国上海崇明东滩，在这里做短暂停留，结识白头鹤、小天鹅等许多新朋友。

During their first long journey southward, baby Great Knots will fly across more than half of the world to reach the tidal flat of Northwest Australia. On the way, they will also visit Chongming Dongtan in Shanghai, China, where their parents have been to, and stop briefly, make new friends such as Hooded Cranes and Tundra Swans.

图书在版编目（CIP）数据

无形的联系：大滨鹬的故事：汉英对照 / 陈婷媛文；吴巍译；王紫绘. -- 上海：上海科学技术出版社，2023.4
（崇明东滩生物多样性科普丛书）
ISBN 978-7-5478-6090-8

Ⅰ. ①无… Ⅱ. ①陈… ②吴… ③王… Ⅲ. ①崇明岛－鸟类－栖息地－沼泽化地－生态恢复－汉、英 Ⅳ. ①Q959.708②P942.513.78

中国国家版本馆CIP数据核字（2023）第038168号

无形的联系——大滨鹬的故事

陈婷媛　文
吴　巍　译
王　紫　绘

上海世纪出版（集团）有限公司
上海科学技术出版社　　出版、发行
（上海市闵行区号景路159弄A座9F-10F）
邮政编码 201101　　www.sstp.cn
上海中华商务联合印刷有限公司印刷
开本 889×1194　1/16　印张 3.5　插页 4
字数 70千字
2023年4月第1版　2023年4月第1次印刷
ISBN 978-7-5478-6090-8/N·256
定价：98.00元

中文版视频

中文版音频

Video in English

Audio in English

本书如有缺页、错装或坏损等严重质量问题，请向工厂联系调换

旗标

全称"迁徙涉禽彩色旗标",是1991年开始由澳大利亚涉禽研究组倡议发展起来的迁徙涉禽专用的鸟类表示系统。即:在鸟类左右腿的不同部位佩戴不同颜色的塑料环片加以标记,迁飞区内每个环志站点的标识方案由相关协议指定并固定使用。旗标通常与传统金属环志一起使用,其优点在于只需通过远距离观察即可确定被标记鸟类的环志地点,极大地方便了对迁徙涉禽的研究。

崇明东滩自然保护区从2003年开始旗标标识工作,其标识方案为右腿上白下黑。自2006年春季开始采用编码旗标,为方便识别编码,将标识方案变更为右腿上黑下白。

Leg Flag Marking

Short for "Migratory Wader Color Leg Flag Marking", is a bird marking system for migratory waders that was initiated by the Australian Wader Study Group in 1991. It means the researchers use different colors of plastic rings or flags on different parts of birds' left and right legs to mark them. The identification scheme of each banding site in the flyway is specified and fixed by relevant agreements. Leg Flag Marking is usually used together with traditional metal ring. Its advantage is that the researchers can recognize the banding site of marked birds by long-distance observation, which greatly facilitates the study of migratory waders.

Chongming Dongtan National Nature Reserve started Leg Flag Marking in 2003. The identification scheme of Chongming Dongtan is: white flag over black flag on the right leg. In the spring of 2006, Chongming Dongtan began to use engraved leg flags. For easy identification of codes, the identification scheme was changed to: black flag over white flag on the right leg.